W9-BGQ-761

FIREWORKS

The Science, the Art, and the Magic

SUSAN KUKLIN

Hyperion Books for Children
New York

ACKNOWLEDGMENTS:
The author would like to thank:
Felix Grucci Jr. and family
Donna and Philip Butler
Phil Grucci and family
The crews and pyrotechnicians who make the fireworks and the volunteers who work the shows
Shea Stadium and the Mets
The town of North Hempstead, N.Y.
The town of Brookhaven and the village of Bellport, N.Y.
New York City Fire Department
New York City Police Department
Marshall Norstein
Robert Schancupp and Fran Katzman Schancupp
Vera B. Williams
Bailey Kuklin
Andrea Cascardi
Joann Hill Lovinski

Printed in Hong Kong by South China Printing Company (1988) Ltd.

First Edition
1 3 5 7 9 10 8 6 4 2

This book is set in 14-point Italia Book.
Designed by Joann Hill Lovinski.

Library of Congress Cataloging-in-Publication Data

Kuklin, Susan
Fireworks/Susan Kuklin—1st ed—p. cm. ISBN 0-7868-0102-6 (trade)—
ISBN 0-7868-2082-9 (lib. bdg.) 1. Fireworks—Juvenile literature. 2. Grucci family—Juvenile literature.
[1. Fireworks. 2. Grucci family.] I. Title. TP300.K84 1996 662'.1—dc20 95-4460

For Concetta Grucci,
with love

Americana

WHHHIZZZEEEE,
WHHHIZZZEEEE, WHHHIZZZEEEE.
A huge blue crysanthemum with red pistils, a cluster of pink and blue peonies, and six long silver tiger tails light up the sky as they dance to music booming from huge speakers. The families picnicking along the beach cry, "Wow!"

Once the show is over and the smoke begins to clear, the audience is still cheering. "I'm happy," smiles Phil Grucci, a member of the family that produced the *pyrotechnic*, or firework, show. "The wind cooperated."

Phil, his grandmother Concetta, his uncle Felix Jr., and his aunt Donna and uncle Phillip are called *pyrotechnicians*, people who make fireworks. Every summer their family and crew members spread out across the country to put on spectacular displays.

When the Gruccis get together they talk fireworks. To describe their work they use words that everybody understands—even their kids. They call fireworks that explode in the upper parts of the sky *aerials*. Aerials blast out of hard paper cases (similar to papier-mâché) that are shaped in spheres or cylinders, called *shells*. Every kind of shell has its very own name, such as a "gold-brocade-to-red-butterfly" and a "green peony with a silver palm core." These shells can be as tiny as three inches in diameter or as big as twelve inches in diameter.

The Gruccis call the explosion of an aerial shell a *break*. Some shells break into squiggly shapes while others, such as the ones they call gold-spangled chrysanthemums, break into fine lines. Some are multibreak shells that burst four, five, even six times, each displaying a different color. Other shells have no color or shape at all, but simply make a loud BANG. These are called *reports* or *salutes*.

oman candles burst from skinny, hard paper tubes that hurtle layers of colored fire up and out into the night. These fill the lower parts of the sky and are known as *illuminations*.

Some illuminations don't shoot up at all. They are called *set pieces*. They're mounted on large frames made of a bamboo grid. A series of pyrotechnic *lances*, small tubes that have color-producing chemicals, are strung out and fused together. When the fuse is lit, a picture appears or a word is spelled with the colored fire.

All the fireworks are stored in forty-foot bunkers called *magazines* that are surrounded by twenty-foot mounds of sand. If one of the magazines should accidentally go off, the sand will prevent it from igniting the others. The tiniest spark can set off the highly explosive chemicals that make up the fireworks.

Even static electricity, which creates those little sparks people sometimes get when they brush their hair, is considered dangerous. Nylon, silk, and polyester easily generate static electricity and must never be worn around fireworks. Employees wear cotton clothing, right down to their underwear. Before entering the shop, Phil touches a copper plate by the door of the building to make certain that no shocks are sticking to his body. Copper is an element that conducts electricity, so any static electricity will harmlessly pass into the plate.

Inside a shop, workers put together the insides of the shells. What goes into the shells and the way they are loaded determines how the fireworks will look. To make a giant flower, Carol loads hundreds of little black balls, called *stars*, in the center of the shell. Some stars are one color. Others are rolled, like jawbreakers, with layers of different color compounds. (Copper salts make blue, strontium salts make red, barium nitrates make green, and sodium makes yellow.) As each layer burns off, a new color appears.

How the Inside of a Comet Works

When the comets burn down to their inner cores, the flash powder ignites and shatters whatever remaining material is left inside. This forces the comet tails to split apart in the shape of crosses. This is different from a star because the star doesn't shatter, it simply burns off.

The golden split comet is the Gruccis' signature aerial shell. Instead of stars, it is made up of thirty-five separate little cylinders called *comets*. It streaks across the sky, leaving a long, fiery tail, like Halley's Comet. As it descends, thousands of golden crosses split off, one on top of the other, until the entire horizon is filled with a lattice of sparkling, glittering crosses.

Those little comets are made of pyrotechnic materials pressed into a pellet with a tiny cavity in the center. Carol fills the comet's cavity with fast-burning flash powder. She squeezes a cap over its top to keep the powder from spilling out.

Inside a six-inch cylindrical case, Carol layers the comets in a circle. The comets are kept in place by materials such as rice or corn.

The Gruccis explode this shell thirty seconds before the grand finale of every single fireworks show. Donna says, "To us, this shell is very special. It is a dedication to my brother who died in a fireworks accident and to my father, who died a few years ago from Alzheimer's disease. They were both so interested in this particular firework. It's our memorial to them. When we fire the split comet, we feel that we are looking up at them. And we know that they are looking down on us."

Once the insides of the shells are completed, the workers wrap them in brown paper soaked in paste. Then they set them out in the sun, where they dry into very hard shells. But they are not ready yet. They still need a way to fly up into the sky.

Cricket, whose real name is Jimmy, is the cousin who supervises this shop. The workers take the hardened shells, attach a *black powder lift charge* at the bottom of each and a *time fuse* at the top, and wrap them in another layer of paper. These charges are connected by another fuse, called a *side fuse*.

Then Cricket inserts a *leader* of fuse into the time fuse. The leader will later be attached to the electrical match. The shell is finished off with aluminum foil to protect it from moisture. Sound complicated? It sure is.

How the Outside Works

The match lights the leader. The leader immediately ignites both the time fuse at the top of the shell and the side fuse. The side fuse shoots a jet of fire down to the black powder lift charge. When the black powder burns, the shell is hurtled out of its mortar, up into the sky. This all happens in milliseconds. When it is at the right height, the time fuse goes off and ignites the black powder bursting charge inside.

It takes two whole days to prepare a twenty-two minute program. Building casings, setting *mortars* (steel or plastic pipes that support the shells and help catapult them into the sky), and running cables and wires take up the first day. The fireworks come later.

At eight o'clock one hot summer morning, a pyrotechnic team drives into a parking lot behind Shea Stadium in New York City. Its truck is loaded with wooden battery sides, mortar pipes, wires, cables, tools, and coolers filled with bottled water, food, and firing command boxes with 300 buttons.

Miles, Kevin, Jeff, Tony, Dennis, and Ed unload the truck and begin building the *batteries*, wooden casings that hold the mortars and fireworks. For this show, "Fireworks Night at the Mets," 222 aerial shells will be held in place by twenty-six battery sections and eighty-four illuminations. The shells end up in groups of seven that will be housed in twelve wooden racks. Miles, the team leader, reviews the *schematic,* a structural diagram of an electrical or mechanical system, that Felix Grucci prepared when he planned the show.

Firing Order

Miles stretches strips of masking tape along one edge of the batteries. He marks two sets of numbers from the schematic with a thick Magic Marker. He circles the numbers that represent the firing order. Next to the circle, he places the numbers that represent the size of the shell and its mortar. For instance, when he looks at the numbers, he knows that position #13 will hold a six-inch mortar and position #123 needs a five-inch mortar.

Everyone pitches in to carry the heavy mortar pipes to their battery berths. Tony places plastic tops on each mortar, and a Mets groundskeeper dumps payloads of sand into the batteries. The sand holds the pipes secure and steady.

It's very hot, ninety-eight degrees, but there is more work to be done. Wires called *webs* are stapled to the batteries. The webs will later connect the shells to the firing panel.

Day two. The truck is chock-full of fireworks. "This is the fun day," laughs Miles as he jumps out of the cab of the truck. "The first thing we do is lay out the shells. Remember when I was writing all those numbers on the tape alongside the battery? Every shell that we unpack has a matching number written on its side. In this way we know exactly where each shell belongs." Kevin walks through the rows and checks that each one is in its right place.

The diameter of a mortar must match the diameter of its shell. If it does not, there won't be enough pressure to catapult the lit firework up into the sky. Here's some easy math: an eight-inch mortar holds an eight-inch shell.

The shells sit patiently on top of their mortars, waiting to be connected to the firing panel and dropped into their tubes. Dennis explains, "These shells are lit by an electrical match. It looks like a simple match head found in a pack of matches, but it's made of metal and attached to a five-foot wire, a leader. Jeff and I attach the leaders to the webs that we worked on yesterday. Later, when one of us pushes buttons to ignite the shells, an electrical current will rush through the wires to the electrical match. That will cause a spark that ignites the fuse in the shell."

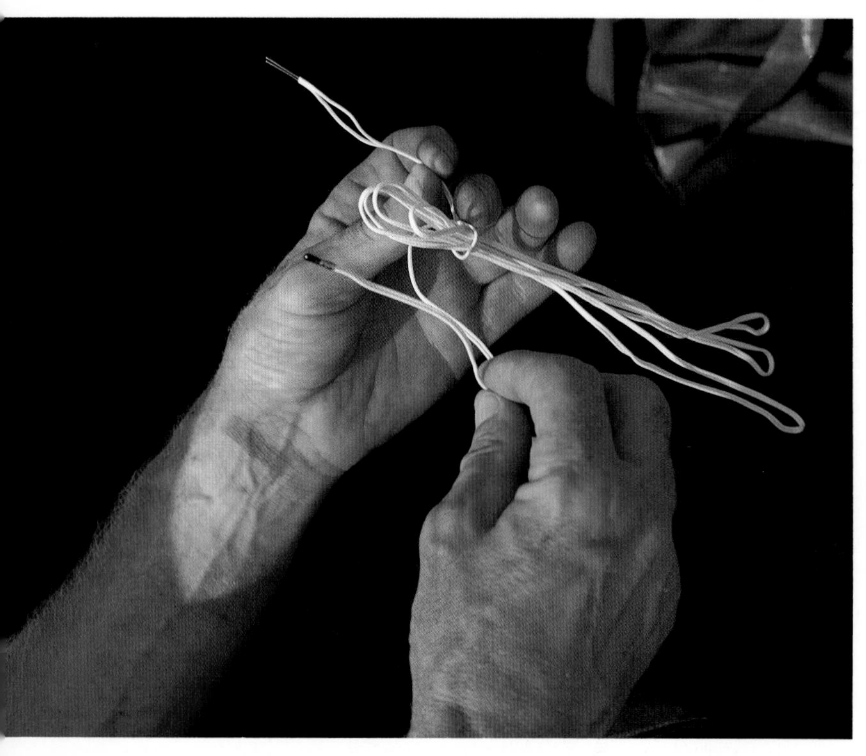

The wires and connections seem complicated, but this team knows its jobs so well they can practically sing the routine:

The firework's connected to the leader.
The leader is connected to the match.
The match is connected to the web.
The web is connected to the cable line.
The cable line's connected to the firing panel.
And the pyrotechnic display is ready to GO.

Miles says, "We're very careful. Everyone here knows what they are doing. We handle fireworks properly, respectfully. We only hold one at a time, we lay everything out to make sure it is where it is supposed to be, we double- and triple-check everything."

The crew members who work the firing station wear hard hats, goggles, and fire-resistant coats to protect them from *fallout*, ash from exploding shells.

This particular show has one firing station. Some have more. For example, the "Fourth of July Star Spangled Spectacular" on Long Island, N.Y., has two stations. One station controls the aerials, and a second one controls the illuminations. Enormous shows, like the "Brooklyn Bridge Centennial," have up to six stations.

Two cars filled with Gruccis arrive for this evening's show. Felix III and his friend are excited to be there. "My father is beginning to teach me a little bit. When a shell shoots up and I don't know the name of it, I'll say, 'Hey, Dad, what's that?' And he tells me. It's neat."

Everyone is calm but Miles. "As it gets closer to show time, I begin to pace. I know that everything is going to work, but I think to myself, 'What if I hit the first button and nothing happens?' After the first shot, I breathe a sigh of relief."

As the sun begins to set and show time looms near, Miles, Ed, and Dennis cover the illuminations with foil to prevent sparks from setting them off prematurely.

During the show the technicians at the firing station wear headsets that tell them when to push the buttons on the command box. They hear cues like these: Fire one! Fire two! The cues are synched with music that accompanies the fireworks.

Earlier, Felix selected the music for the show. He decided the exact time that the fireworks would explode to the music. He can match all kinds of fireworks when he choreographs a song.

Felix says, "I get the feel of the music and then I see the visual effects pyrotechnically. I hear a high note and I imagine a big burst of color. Then, when a low note comes along, I add a soft color, like blue. In a way I'm like a choreographer who creates a ballet, only I use fire in place of dancers."

Two soaring palms.

Blue and golden spider.

Silver tourbillion with two chrysanthemums.

Blossom after thunder.

"Back up, everybody. We're bringing in the truck," shouts Police Officer DeScalo. "We don't want any accidents."

"Yo, Anthony, you made it," welcomes Tony as he opens the back of the truck. Anthony makes all the set pieces. He and his crew bring two huge fireworks signs that say "Mets" and "Coke." Beyond the outfield the crew puts together large scaffolds and attaches Anthony's signs.

Whichever Grucci runs the show is responsible for the entire production. Tonight it's Felix's turn. The buttons must be pressed in the correct order. The music must be in sync with the fireworks. The safety zone must be secure. And, most important, the crew and the audience must be safe.

"Is the wind coming up? Are those rain clouds?" asks Felix of no one in particular. There's less than an hour to go.

The ball game is over, and the lights in the stadium go out. Felix talks to Miles through the headset. "Whenever you're ready, Miles, let's do a practice shot." WHOOSH . . . Jeff shoots a shell off into the night.

"Oooooh," shouts the crowd, not realizing that the show has not yet begun.

"That looked good," Felix reports. The audience is quiet. Felix asks Miles, "Is everyone in position?"

"Roger."

"Okay. We go in thirty seconds."

"Have a safe one," whispers Donna. Music plays. **FIRE ONE!**

A gold-spangled chrysanthemum with a red magnesium comet explodes in the sky!

FIRE TWO! PSHSSSSSSSSTU! Roman candles race toward the comets.

FIRE THREE! WHHHIZZZEEEE! More Roman candles . . .

FIRE THIRTY-THREE! A blue spider turning red, then gold, lingers in midair . . .

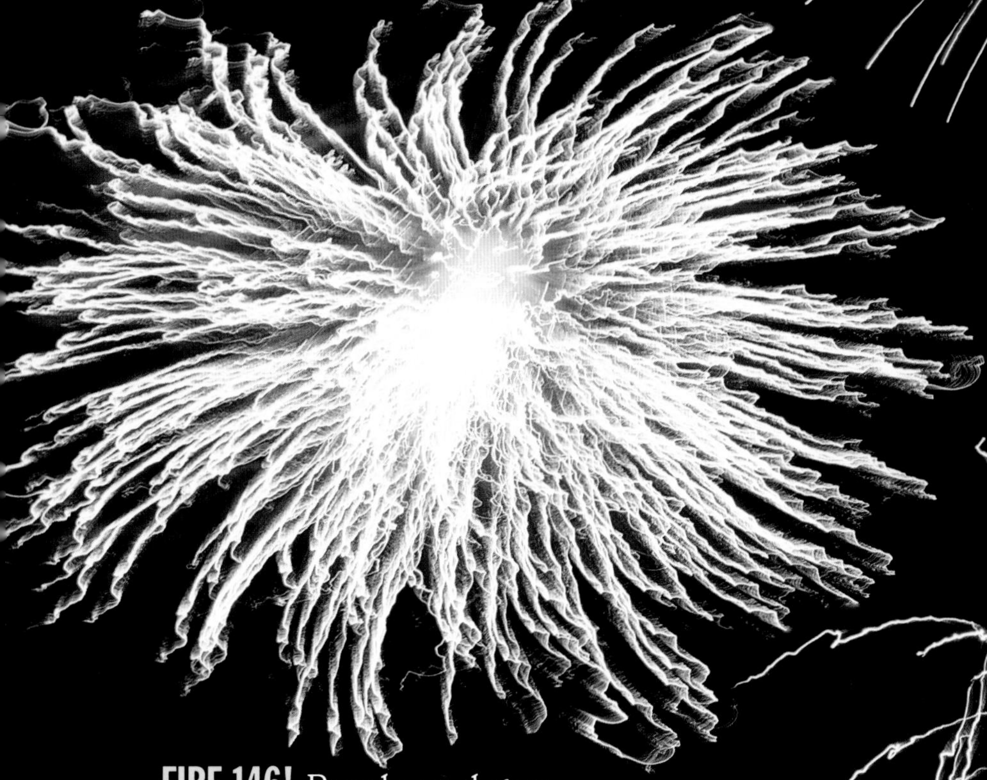

FIRE 145! A white wave!

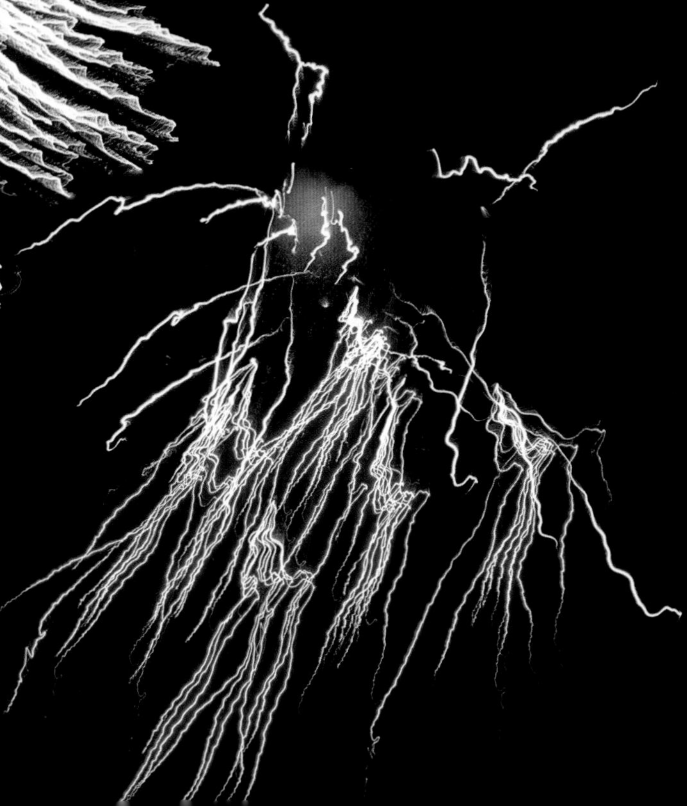

FIRE 146! Purple and green jumping stars!

For five seconds the sky is quiet. And then **BOOM!** The golden split comet lights up the sky exactly thirty seconds before the grand finale. "Ah-h-h-h-h," cheers the crowd.

BANG-BANG-BANG-BANG-BANG. Jeff pushes one button, and the entire sky fills with color and noise.

Eight Roman candles attached to white aerials with silver tiger tails reach up toward gold-spangled chrysanthemums. A titanium flash salutes BOOM in the sky. And another BOOM. And another. Sparkling red chrysanthemums POP, POP, POP as they follow color and titanium thunder. The audience applauds wildly as the smoke clears.

Moments after the show Felix leans back and breathes a sigh of relief. "Beautiful job! This show went off without a hitch." A big smile fills his face.

Concetta Grucci is always nearby to congratulate her family. "I'm so proud of my children," she says.

Donna says, "Every year we try to get better and better. It's a challenge to make it more exciting, more thrilling."

The Gruccis attribute their success to their working together as a family. It's hard work, but in the end, they get pleasure out of bringing these dazzling shows to the public. Felix says, "Success can be measured in a lot of different ways. We measure ours in the happiness that we are able to bring to people. If we put a value to it, I'd say we are probably among the richest people in the world."

ABOUT FIREWORKS:

Fireworks are the heart of such festivals as the Fourth of July, New Year's Eve, Mardi Gras, and the Chinese New Year. The earliest known fireworks were projectiles, similar to Roman candles, that spit balls of fire from bamboo tubes. They were most likely created during the tenth century in China, where they were used in ceremonial events.

Black powder determined the speed, height, and bursting power of the shell. Soon the black powder and projectiles were used as instruments of war, first attached to arrows, then as gunshot and cannonballs.

In 1242 English monk Roger Bacon wrote down the instructions for the preparation of black powder. By 1353 the Arabs had developed this formula to make the first gun, a bamboo tube reinforced with iron to withstand the force of the powder. King Louis XIV of France (1643–1715) returned to the ceremonial use of fireworks for their exquisite visual effects at his lavish galas and dances at his palace at Versailles.

The Italians are credited for developing colored fire—reds, blues, and greens—for displays to commemorate religious festivals.

Most of today's fireworks come from well-established, family-owned businesses. They guard their black powder recipes and mixing procedures as trade secrets that are passed on from one generation to the next.

ABOUT THE GRUCCI FAMILY:

Donna Grucci Butler says, "We bring different ideas, different personalities, and different generations to our business. But the one thing that keeps our family strong is that we never lose sight of who we are and where we came from."

In 1850, Angelo Lanzetta founded a pyrotechnic company in Bari, Italy. Twenty years later, he immigrated to America and set up shop on Long Island, N.Y. After Angelo's death in 1899, his son Anthony carried on the family business. In 1923, Anthony invited his nephew Felix Grucci to become an apprentice.

Anthony died in 1938, and Felix took over the business. To make ends meet, Felix worked in a club as a drummer. There he met Concetta Didio. They were married in 1940 and had three children: James, Donna, and Felix.

Felix Sr. says, "Some kids hung around the delicatessen because their mother and father made deli foods. Ours made fireworks."

Over the next two decades Felix Grucci Sr. gained the reputation as a master of his art. With his brother Joseph he invented the stringless shell to eliminate fallout and created the golden split comet. Felix Sr. choreographed fireworks to music, and made an atomic device simulator for the Department of Defense to train U.S. troops.

Felix Sr. and Concetta took their children to every program. "We did everything as a family," explains Donna. "We traveled together to all the shows. We felt proud that we brought so much joy and entertainment to so many people."

Felix Jr., Donna, and Jimmy grew up, got married, and had children of their own. In keeping with tradition, they joined the family business and took their children to all the shows. The business flourished.

Then tragedy struck. In November 1983 an accidental explosion destroyed the Grucci building complex and killed two family members—Jimmy and his cousin.

Phil says, "After that, we had no property, no building, nothing to work in. My father and our cousin were dead. We sat around my grandparents' dining-room table and just stared at each other. We were all we had left."

One week after the accident, Felix Sr. held a family meeting to decide what to do. Should they go on? Could they go on? Then the letters started pouring in.

Fans wrote about the joy the Gruccis' fireworks displays brought to their lives and begged them not to give up. By 1986 they were back in business. And now Phil (Jimmy's son) and his wife, Debbie, take their children, Lauren and Christopher, to watch their dad work.

Along with such annual events as numerous Fourth of July celebrations, the Gruccis produced the fireworks displays at the Lake Placid 1980 Winter Olympics, the 1982 World's Fair in Knoxville, the 1983 Brooklyn Bridge Centennial, the 1986 Statue of Liberty Centennial, and the inaugurations of presidents Reagan (1981, 1985), Bush (1989), and Clinton (1993). In 1979, the Gruccis had become the first American family to win the gold medal at the Monte Carlo International Fireworks Competition. At that time the press dubbed them the First Family of Fireworks. They still are.